A Green Reef

The Impact of Climate Change

Cover design: Debbie Geltner
Book design: WildElement.ca
Author photo: Erika de Vasconcelos

Library and Archives Canada Cataloguing in Publication

Henighan, Stephen, 1960-, author
    A green reef : the impact of climate change / Stephen Henighan

Issued in print and electronic formats.

ISBN 978-1-927535-27-1 (pbk.).--ISBN 978-1-927535-28-8 (epub).--
ISBN 978-1-927535-29-5 (mobi).--ISBN 978-1-927535-30-1 (pdf)

    1. Climatic changes. 2. Climate and civilization. 3. Nature--Effect
of human beings on. I. Title.

  QC903.H44 2013   304.2'5   C2013-902014-4
                                                   C2013-902015-2

Legal Deposit Library and Archives Canada
et Bibliothèque et archives nationales du Québec

Linda Leith Publishing acknowledges the support of the
Canada Council for the Arts.

Printed and bound in Canada by Marquis Book Printing Inc.

Linda Leith Publishing
P.O. Box 322, Station Victoria
Westmount QC H3Z 2V8 Canada
www.lindaleith.com

CHAPTER ONE
THE END OF ABUNDANCE

On the West Coast of Canada, natural abundance feels inevitable. The gigantic scale on which nature surrounds us persuades the observer that none of this could ever change: the trees, inlets, and animals seem to belong not to an evolving life-process, but to a fixed, postcard-like idea of grandeur. Yet, once we look closer, this world feels fragile. Behind the roadside ribbon of decency, maintained in order not to shock the public, stretch the gutted stumps of the clearcuts. On a still morning, the gossamer mist breathed by the mysterious world of a northern rainforest mingles with the fumes that belch from a logging truck. A black bear ambles into the bush at the sound of a chainsaw. The eagles that used to whirl in the sky are scarce. Staring into the empty air, one thinks that if this place, where nature is more generous than anywhere else in the northern Americas, can turn bleak and stunted, no haven is secure.

I was introduced to the West Coast in 1984. Having finished university, I drove across the continent, sampling landscapes and cities, until I reached California. From there, there was nowhere to go but north. I turned in my Drive-Away car in Oakland and rode by bus to Oregon, Washington and British Columbia. I stayed with relatives in Victoria. On Canada Day I stared out past the exploding fireworks at the snowy peaks of the Olympic Mountains, across the Juan de Fuca Strait in Washington State. The shadowy cushion of snow, glimmering high in the sky on the clear July first evening, lay to the south of me, yet it beckoned me north.

Four days later I caught a 6 A.M. bus out of Victoria. It was a local bus as far as Nanaimo, where it became an express to Port Hardy on the Queen Charlotte Strait, near the top of Vancouver Island. This eight-hour trip was my introduction to the natural world of the West Coast. The journey began with a long, slow thrash through early-morning fog and dense firs. As the fog cleared, it unveiled foliage of a denser, darker green: a landscape of rainforest profusion and Nordic sturdiness. I thrilled at the passing glimpses of snow-dented peaks as the bus began to climb. Yet here, too, I received my first jolt. Many of the hillsides had been logged out and clear-cut. Nothing in the quaint streets and cultivated gardens of Victoria had prepared me for this careless destruction. The bus climbed almost imperceptibly to an ear-squeezing altitude. We descended past pink-orange rock facings, with shrubbery clinging to each fissure in the rock. The bus emerged into a landscape of gentler hillsides. Here, too, many of the slopes had been logged.

The visitor's first encounter with the natural wealth of the West Coast is also an encounter with the threat to that wealth. In a place where people, though more numerous than they appear at first, are far less prevalent than trees or wild mammals, the pressure of human populations makes a glaring impact. The scramble for resources to satisfy the planet's ever-multiplying hordes of humans stands out most starkly, as though by contrast, in areas where humans do not predominate. My trip north that summer became an apprenticeship in the idea – a novel one to me at the time – that nature, that ferocious tyrant who had dictated the lines of my life through the howling winters of my Central Canadian childhood, was in retreat before the demands of throngs of people.

I had arranged to meet a friend in Alaska. Today that friend is a prominent plant and wildlife biologist; then he was a graduate student, beginning his research. As I travelled north, I felt saturated by the natural world, yet penetrated by a keening melancholy. In Port Hardy I caught a ferry to Prince Rupert, where the MV *Malaspina*, destination Skagway, Alaska, waited at the next wharf. Down the quay, an enormous Japanese tanker was being washed down. Prince Rupert, a day closer to Japan than Vancouver, was already preparing to become the conduit for shipping Canada's raw materials to Asia. From here on, my experience of the natural world of the West Coast was filtered through the light of days that got longer and longer until they were nearly endless. At first these days were grey, in an infinite variety of muted pastel-like tones of glowing dimness. The towns where the *Malaspina* docked received more than 320 days of rain a year; in each outpost there was a wash of ramshackle wooden buildings along the waterfront, with the liquor store right in the middle. I had never felt more aware of the elements than I did sleeping under an overhang on the deck. The Inside Passage was cool, the water flat. A grey light filtered through the banks of permanent cloud from 5 a.m. to 10 p.m. with little change in intensity. The greyness waned into night. Jumbled hills, dark beneath the cloud, covered the foreground with spruce and hemlock stacked branch-to-branch for hundreds of kilometres. The mountains behind them might as well not have existed; seemed to exist only on maps. The sun was a rumour, the stillness enormous.

My friend, who lived in Anchorage, had driven two days to Haines, in the Alaska Panhandle to pick me up. From here we would drive over the Chilkat Pass, across a corner of British Columbia that had no road link to the rest of the province, into the Yukon, then back to Alaska, where we would spend a month hiking. As the ferry approached Haines, the cloud lifted. To the east, cliffs covered with spruce and cedars, seamed with waterfalls, plunged into the ocean, lingering cloud sliced off their tops. On the left black mountains, patched with snow, shot upward through the cloud. One of them cradled a glacier. From the deck, I saw my friend getting out of his battered Volkswagen Beetle.

I got a better sense of the landscape from the passenger seat of the Beetle than I had from the ferry: the swampy murk, bush and low hills in the foreground, the sheer *complication* of the land that culminated in the sought-after, often-concealed mountain peaks. During the next month, with the help of the Beetle, we went to lookouts where I saw glaciers floating in a bay, trails where we would hike up mountainsides to tread glacial moraines, feel the cold radiating off white-blue glaciers ruffled like frozen coral, then walk across them. We would camp in high, drizzly passes and ford icy rivers in bright sunlight. I saw dozens of caribou, foxes, and mountain goats, and squinted at ambling grizzly bears through binoculars. I watched the sun set at 1 A.M., skid along the horizon and swing up again a little more than an hour later. I had never before been in the presence of a landscape so vast and raw. I had never explored a place that seemed to be for animals and wilderness rather than people. Central Alaska did not have the quasi-tropical abundance of the British Columbia coastline. Its thin tundra and long patches of bare earth were a reminder that it would be all too easy for people to ruin this place. One afternoon, in Denali Park, we climbed to the top of a high spur from where we were able to watch a herd of caribou passing right below us. In the distance (but not too distant), a grizzly bear prowled on all fours along the crest of a ridge, surveying the open landscape in front of him. I took a backward step. My boot landed on a patch of crinkly green, not much larger than the palm of my hand, that spread over the brown earth.

"Careful," my friend said. "That lichen took five hundred years to grow to that size."

In Alaska my awareness of the thoughtless wrecking of a habitat presumed to be infinitely renewable, awakened by my trip up the British Columbian coast, fused with the realization that not all of

4

nature shared the coast's abundance. In B.C. you could clear-cut and lots of trees remained – or so it seemed in 1984. Alaska, for all its magnificent landscapes, did not offer the comfort of this illusion. The sub-Arctic was a more austere environment. It raised questions about how people and nature would coexist. Wrestling with whether to go to graduate school in September, I wrote in a red exercise book in the tent every evening, trying to figure out what to do with my life. The biologist asked me about my writing; his girlfriend, who would be coming up from Oregon later in the summer, also wrote. "I would write," he said, waving his hand, "if I could show people exactly what a black spruce looks like." He found other concerns more pressing. Like what? I asked. "Like what's going to happen to all this in the future."

Having spent my travels worrying about my own future, I had overlooked the prospects for humanity. Did the biologist think we would survive? The question sounded melodramatic. In those dying days of the Cold War, informed people who worried about our survival read Jonathan Schell's *The Fate of the Earth* and feared nuclear war, rather than the consequences of our sacking of the natural world. The biologist did the math for me, charting the exponential increase in the planet's human inhabitants. There would be six billion people in the world by the year 2000, perhaps nine or ten billion by 2050.... But it would probably all be over by then. We would be choking on fumes, murdering each other for the last scraps of food and mouthfuls of fresh water. Why? "Too many goddamn people."

In 1984, only scientists voiced such possibilities, usually among themselves. When I thought about global issues, my attention turned to the civil wars in Central America, not to overpopulation or the environment. The possibility that human civilization might end other than by nuclear war was new to me. The biologist's conclusions were startling; at the same time, the mood that had engulfed me during my trip had rendered me susceptible to this outlook.

We were sitting on a high, rocky outcrop. In the sunlight, we could see for dozens of kilometres, all the way to the sharp flanks of the mountains. Above these bare mountainsides, snow-covered peaks butted the sky. Nowhere in this vista was there evidence of human life.

The biologist repeated his words under his breath. "Too many goddamn people."

CHAPTER TWO
NO OTHER WORLD

In 2011 the United States launched its last space shuttle mission to the International Space Station (ISS). After thirty years and 135 launches, the shuttle was discarded as expensive, unreliable, and unable to recoup on the hundreds of billions of dollars that had been invested in its development and maintenance. The closure of the ISS was announced for 2020. The Russians continued to send cosmonauts into orbit close to the earth, the Chinese, having launched the first taikonauts, announced vague ambitions to visit the moon by the 2030s, yet it was obvious that intergalactic travel had slipped down the list of planetary priorities.

To anyone raised during the Cold War, when expanding the realm of human activity to other planets, and eventually other solar systems, seemed to be only a matter of time, this news had a certain poignancy. How many of us gathered around black-and-white television sets as Neil Armstrong's shadowy boot descended from a ladder onto the lunar surface! We followed *Star Trek* and gobbled down science fiction novels. Just as it never occurred to me, until I was perched on a high bluff in the Alaskan wilderness, that human civilization was in peril, I never doubted that I would live to see the day of free and easy transportation to the Moon, Mars, and beyond. Yet, like most people, I failed to realize that my conception of the future was a product of the ideology and circumstances of my past.

The frenzied race to get to the Moon by the end of the 1960s, announced by US President John F. Kennedy, was propelled by Cold War obsessions: the need to recover from the loss of face caused by the Sputnik launch and establish control of the heavens in order to confirm the superiority of our side's system here on Earth.

It is astonishing that, with the relatively primitive technology of the 1960s, men were able to land on the Moon. In fact, to the average Blackberry-wielding yuppie of the 2010s, the claim that the rudimentary computers of 1969 were able to chart a spaceship's route to the Moon is downright suspicious. And if it was possible to go to the Moon in 1969, why did President Richard Nixon abolish the lunar program in 1973? Surely with time the trip should have become easier? Witnessing the multiple failures that have attended NASA's attempts to use the technology of the early twenty-first century to supply and maintain the ISS, which circles in the relative proximity of the Earth's orbit, has stoked many people's secret belief that the Cubans are right to teach schoolchildren the lunar landing was faked for propaganda purposes in the Arizona desert. Whether or not this is the case, the idea of space as a boundless "final frontier," as *Star Trek* proclaimed, has withered. To offset the ensuing pessimism, the U.S. government followed up the final shuttle launch with the announcement of a century-long feasibility study on intergalactic travel. Given that the select group of scientists chosen for the "One-Hundred-Year Starship Study" is not due to deliver their report until approximately 2115, it is clear that not only will we not be building cities on Mars any time soon, it is possible that humans will never go there. Unmanned craft have visited all of the planets, yet whether humans would survive the return journey of more than a year's duration to the red planet, with its attendant exposure to solar flares and other perils of the intergalactic highway, is a very large question. Even the Moon seems increasingly irrelevant to our dilemmas here on Earth. As our planet's orbit becomes clogged with satellites conducting cellphone signals, predicting the weather or spying on the competition, it grows increasingly evident that this hub of activity is the outer rim of our world.

We have no planet but this one. For much of the last fifty years we have been able to delude ourselves on this point. But with the elimination of an active space program, underscored by the final shuttle launch in 2011, this evasion ceased to be viable.

As we face the prospect of "too many goddamn people," with its calamitous side-effect – which my friend and I did not discuss in detail in 1984 – of severe climate change, the resources we can muster to solve the problem are limited to those present here on Earth. Since I first began to worry about climate change a few years ago, I've been astonished by the number of people who have shrugged off my concerns by casually saying, "There's no problem, we'll just move on to Mars," or "Don't worry, if the US Midwest turns into a desert, we can always build greenhouses on the Moon and produce more food than we'll ever need." Fifty years of television science fiction, the lightweight sidekick of Cold War propaganda, has left us deeply misinformed about our place in the universe. We will never be interstellar cowboys. Like survivors of a shipwreck, humanity clings to a small, fertile blue-green reef in the middle of a limitless, hostile ocean. If we become too numerous to live on the bounty of this reef, there is nowhere else to go. We will die gasping for breath.

The end of the US space program, and the comparative lack of urgency with which the Russians and Chinese approach their space endeavours, is a salutary event: a reminder, precisely when we needed it most, that human life is inseparable from the life of this planet. A renewed commitment to Earth, a sensitivity to the interconnections between our ability to breathe and nourish ourselves, and those of the plants and animals, is indispensable to us. There is no escape, no other place of refuge. We live here or we die here. At the moment, the odds are that a large number of human communities, probably a majority of them, will die. Our death will be caused by the fact that more and more human beings aspire to live – and now have the means to live – the lives enjoyed by most of the people who will have the opportunity to read this book: eating fruit in the dead of winter, flying to other cities for business or to other continents for pleasure, turning on the air conditioning when it gets too hot, owning a private car and using it even for errands in the town or city where we live. This life posed only a limited threat to the home we all inhabit as long as it was restricted to the privileged, and not excessively heavily populated, regions of Western Europe and North America. As this life – or "lifestyle," as it is now called – has spread throughout the world, particularly to very heavily populated nations in Asia, it has become a liability. And, since it has spread hand-in-hand with an ideology of mass consumption, profit maximization, and individual liberty, to the verge of libertarian social irresponsibility, it has become impossible to curb.

We are victims of our global triumph. There is no natural habitat on Earth where we do not dominate. A species like any other, we measure our success by our proliferation, our ability to flood the world with genetic extensions of ourselves until there are "too many goddamn people." Yet we are a species like no other. Where other species consume according to instinct, eating sprouts or insects or fish or other mammals, we consume omnivorously and extravagantly. We gobble down not only plants and animals, but trees, rocks and petroleum. We simply must sample caviar or prawn eggs or Pacific salmon or ostrich or sushi from the ocean floor, regardless of the consequences. Over even brief periods of time, we have grown bigger and fatter and needier. North American airline regulations, written in the 1970s, assumed an average airline passenger weight of 170 pounds (77 kilograms); by 2010, the regulations were out of date because the average airline passenger weighed 215 pounds (98 kilograms).

To fuel our pleasures and stuff our expanding guts with New Zealand lamb, South African apples, Chilean wine, and thousands of burgers made from beef raised in clear-cuts in the former Amazonian rainforest, to give our love a rose that was picked yesterday in Kenya or Ecuador, we burn carbon, deplete the ozone layer, and heat up the planet. We are the most savage of all animals because, unlike other predators, we do not eat individual members of other species, but rather drive the entire species to extinction. We are aware of how our species is doing on other continents: we know that Africa, a dismal place for most of its inhabitants as recently as the mid-1990s, is now showing promise. Millions of Africans remain trapped in desolate poverty, or are even vulnerable to periodic bouts of starvation; yet the number of Africans who can live like us, and consume like us, increases every year. By speeding up the loss of species and raising the temperature of the planet, this victory for humankind brings our own species one step closer to self-destruction.

Human activity has accelerated extinction. It is a normal part of the evolutionary process for superannuated species to go extinct. Prior to human interference, about one species out of every hundred thousand went extinct each year. This rate has increased by a factor of at least one hundred and perhaps as much as one thousand (not all species have been documented, and scientists tend to be cautious about declaring species extinct); species are now going extinct before successor species have been able to develop. The last time species were disappearing as quickly as they are today was sixty-five million years ago, during the extinction of the dinosaurs. Ironically, our primate cousins, who look and often act in ways that have an eerie resemblance to our own physiognomies and behaviour, are among those most endangered by the fall-out from our gluttony. African economic growth – as we conceive "growth" – reduces the chances of survival for Kirk's red colobus monkey or the mountain gorillas of Uganda and Rwanda, just as, in earlier eras, economic growth put an end to Great Britain's wolf populations and North Americans mowed down the buffalo and clubbed the passenger pigeon.

In fact, our history of spreading extinction goes much farther back; it is almost our defining characteristic to snuff out other species. The appearance of humans during the Pleistocene epoch (the time between 2.6 million and 11,700 years ago) coincides with the great Quaternary Extinctions, a rolling succession of species annihilations that started about fifty thousand years ago and in many cases coincided with the arrival of humans. As people walked out of Africa, where we originated, and invaded the rest of the planet, many of the world's populations of larger mammals were hunted to their last survivor. In North America, the crescendo of this march of death occurred between ten and thirteen thousand years ago. Of the one hundred and three species of large mammals that became extinct during this time, seventy-nine were found in the Americas, the most recent territory conquered by humans. In Africa, where large mammals had been coexisting with humans for far longer (and, some paleontologists think, had evolved to survive in human-dominated environments, much as the raccoon has done in urban areas in recent decades) only two large mammal species became extinct during these centuries.

The holocaust wreaked by the new human inhabitants of the Americas is one of the central events of pre-history. Prior to the Quaternary Extinction, the plains, forests and mountains of what is today Canada harboured mammoths, mastodons, sabre-toothed cats, pronghorns, stag moose, Yukon wild asses, northern llamas, native American horses, short-faced bears, and dozens of other creatures that fell before the thoughtless hunters' clubs, spears, and arrows. Today we praise the aboriginal peoples of the Americas for having learned to live in harmony with nature. This wisdom was attained at a price. By the time Europeans began to settle in the Americas in the early sixteenth century, indigenous people had enshrined animals as sacred. The hunt was a religious ritual, con-

ducted with respect and restraint to ensure that species would survive to provide sustenance in coming years. At some point thousands
of years ago, the indigenous peoples absorbed the knowledge that by
depleting the environment that sustained them, they would also de-
stroy themselves. But this awareness was not there at the beginning
of aboriginal American civilization, just as it is lacking in the ways in
which we comport and govern ourselves today.

Like the aboriginal people of the Late Pleistocene, we are on the
brink of demolishing our own longhouse. Yet, even as species vanish
before the encroachment of highways, oil fields, clear-cuts, housing
developments, poachers, the drying-up of wetlands, and the blight-
ing of coral reefs, others migrate in response to the threat of climate
change. Just as in Central Canada, the raccoon has adapted to the
human environment by changing from an animal of the forests to a
creature of the city streets – masked neighbours who hang improb-
ably from lamp-posts in downtown Toronto and feed off the city's
garbage as they once fed off the residues of the woods – so tropical
life-forms that were not part of the Canadian landscape in the past
are venturing farther north as the planet warms. The mountain pine
beetle, the Japanese ladybug, and the African killer bee, previously
contained by our long, cold winters, contribute to the alteration of
our environment. As our world grows hotter, the range of some insects and animals expands, while others, like polar bears, struggle
to survive. The extinction of species proceeds along multiple axes,
sometimes as the direct result of human intervention, and in other
cases as collateral damage of the larger process of global warming.
Many species' best hope for survival lies in the possibility that the
species that caused our planet to begin getting hotter, the species that
destroys the habitats of others, will have its numbers drastically re-
duced.

And that is almost certainly about to happen.

13

CHAPTER THREE
THE PROGNOSIS

"I am increasingly finding that at elite dinner parties there is already discussion of who the survivors will be."

The words are those of James Martin, a British multimillionaire who gave away part of the fortune he made in computers and textbook publishing to found an institute at the University of Oxford dedicated to the search for solutions to the problems of the twenty-first century. We will not be able to predict how many people or nations will survive until we know how high the global temperature will rise. On our densely built-up planet, a rise of even one degree Celsius will be calamitous. That one degree is a global average; since two-thirds of Earth's surface is covered by oceans, which will remain cooler, a one-degree average rise means almost two degrees over land, sufficient to transform the contours of our world in terrifying ways, and possibly as high as four or five degrees at the poles, where estimates of the speed with which the icecaps are melting are constantly being revised upward. Countries that take science seriously are acting: India is building an enormous fence to keep out the tens of millions of refugees it expects when low-lying Bangladesh is flooded; China is securing oil and mineral reserves in Africa, and buying up swathes of land that is expected to be arable when the temperature rises, in places such as Patagonia and central Quebec. The United States, whose citizens have the largest per capita carbon footprint in the world –almost ten times that of the average citizen of China – has been prevented from moving forward by a poisonous blend of fundamentalist religious dogma, which dictates that humans do not control their destiny, anti-tax rhetoric, which makes the suggestion of a carbon tax treasonous, and the puppeteer-like control exercised by oil and armaments firms on the words of US politicians. "Does a top US senator really believe it when he says that global warming is a hoax, and that all thousand top scientists in the International Panel on Climate Control are liars, and that they all tell the same, meticulously detailed lie?" James Martin asks. The suicidal subservience of American politicians to corporate suzerainty is revolting; yet in the short to medium run, it barely matters. The process of global warming has advanced too far be stopped. Disaster is inevitable. The only question is the severity of the blow.

Taking 1990 temperatures as a baseline, a rise of a little more than one degree Celsius by 2040 will result in the flooding of most coastal areas of South-East Asia, the virtual obliteration of agriculture in Mexico, Central America, the Caribbean, and North Africa, and prolonged drought in southern Europe and the southwestern United States. Yet, at this stage, a rise of between one and two degrees is scarcely credible. The processes that we have unleashed will almost certainly drive the temperature higher. If all goes well, if we are lucky, the rise in temperature by 2040 may be limited to 2.6 degrees. But even at this level, the world of 2040 will be unrecognizable. A rise of 2.6 degrees over the 1990 baseline (which means upwards of four degrees over land) will melt ice sheets at both poles. New York, Boston, London, Shanghai, Bangkok, Cairo, Buenos Aires, Lisbon, Mumbai, and many other important coastal or riverside cities will be flooded, adding tens of millions of new refugees to those already scrambling for homes and scarce resources. Harvests will plummet around the world and agriculture will no longer be viable in places with climates we would classify today as sub-tropical.

In addition to winding down much of the mass production of goods and services that we take for granted, this environment will cause the disintegration of representative political systems.

In large countries plagued by environmental disasters in different corners of the land, central governments may lose the ability to provide disaster relief and simply throw outlying regions back on the efforts of their own citizenry. This will deprive national governments of credibility and favour the rise of regional strong men. Democratic procedures will be hobbled by a faltering supply of information as electricity generation becomes unreliable, converting internet access into a privilege of elites. The world's two most dangerous military powers, China and the United States, will both begin to run out of water and turn their eyes northward. China is very likely to invade Siberia in order to take possession of the vast freshwater reserves of the Lake Baikal system, without which much of its population will die of thirst. The United States will either induce Canada to hand over its fresh water – 9% of the world's total – through diplomatic pressure, or will subject our country to military occupation. Should the U.S. government have disintegrated to a point where it is unable to coordinate this sort of military action, individual Americans, as has been their historical tradition, may form militias to take matters into their own hands. In Africa, the Middle East, Central Asia and parts of Latin America, droughts will kill tens of millions of people. Northern Europe – Scandinavia, parts of Germany, Poland, Scotland, and European Russia – may remain agriculturally self-sufficient; if this is the case, these will be fearful, militaristic societies walled in against desperate refugees fleeing from countries like Spain, which will have run out of water.

Even these circumstances are mild by comparison with the world that will be wrought if current predictions are correct that rising temperatures will cause the permafrost that underlies one-quarter of the northern hemisphere to melt, releasing huge quantities of methane gas and carbon dioxide that will send global warming into overdrive. If this happens, temperatures may rise to six, or even eight or nine, degrees above the 1990 baseline. In a kind of brutal cosmic justice, we will have created our own Quaternary Extinction, thrusting upon ourselves, on a global scale, the species-wide destruction that humans visited on large North American mammals thousands of years ago. At nine degrees above 1990 baseline, we will recreate the Eocene epoch, the hottest time in Earth's history. But that was fifty-five million years ago, when there were no humans on the planet. A new Eocene period would concentrate human habitation around the fringes of the Arctic Ocean; human communities might also survive in mountainous areas where forests and rainfall persisted, such as Patagonia, New Zealand, the Scottish highlands, and parts of British Columbia. It is unlikely that the planet's total human population would exceed a few hundred million. The sea level would have risen by seventy metres, shrinking the continents. Over much of the Earth's surface, daytime temperatures of 50 to 60 degrees Celsius would be routine. The oceans, clogged with bacteria, would be an oily purple colour and would smell like rotten eggs. The sky, filled with sulfurous bacteria bubbling up out of the torpid waves, would be not blue but green.

None of these scenarios is certain, yet some variation of one of them will occur. It is too late to rein in the forces we have released. Good policy, which addresses the environment as humanity's most urgent priority, will produce a better outcome than half-hearted ameliorative measures, or pretending that the problem doesn't exist. Yet, at this late stage, no policy can ward off cataclysmic change. We go about our daily lives knowing that the world we inhabit cannot exist for much longer.

CHAPTER FOUR
PHONY WAR

On September 1, 1939 the German Army invaded Poland and the realization spread through most of the northern hemisphere that the world was once again at war. On December 10, the First Division of the Canadian Army sailed for Europe. The Canadian troops expected to enter a growing conflagration, but when they reached their bases in the United Kingdom they found that the Allied and Axis sides, armed and mobilized, were at war in word but not in deed. For months, nothing happened. As some wag put it, having expected to confront the German Blitzkrieg, they found themselves in a Sitzkrieg. Known as the Phony War, this period of unearthly calm lasted until April 1940. The knowledge that death and destruction were on the horizon did not prevent many people in Europe from continuing to live as they had before. My grandparents, for example, decided during this lull before the storm that they would have their fourth child, who twenty years later would become my mother. My grandfather continued to work at the clothing shop he owned in London until the day he went to work to find that German bombers had left a large hole in the ground where the shop had stood. The awareness of approaching disaster did not alter my grandparents' behaviour. Only the next spring, when Germany invaded Norway, did the full import of their decision to enlarge their family become apparent.

Today we are once again in a Phony War. This time the antagonist is the damage we have done to our climate. Most people who are attentive to the news media are aware of the irrefutable evidence that the planet is becoming warmer as a result of human activity. This conclusion may not be universally accepted in Fort McMurray, or on George Bush's ranch; but, beyond these outposts of obscurantism, the debate is over. We know that life-altering, and possibly cataclysmic, change is coming, and we continue to live as we have always done, burning as much fossil fuel as our incomes permit. We justify ourselves by telling friends how we recycle newspapers, use low-energy light bulbs, eschew bottled water or bring cloth bags to the supermarket. My own claim to environmental virtue is that I have never owned a car; this pretension is nullified by my habit of making long trips on airplanes half a dozen times a year. Our small gestures towards environmental responsibility, which might be significant in the context of a large-scale effort to decarbonize civilization, are rendered meaningless by a society that fails to address the central issue of people in wealthy countries consuming resources at a rate that guarantees that within twenty to thirty years, people will begin to starve.

Once this consciousness creeps into your head, it never goes away. No act is innocent, no moment of triumph untainted by the apocalypse that lies ahead. I sit in a committee meeting and listen to an academic vice-president describe how the "competitiveness" of the university where I work depends on expanding internationalization. As I take notes on his plans to send ever larger numbers of students on semesters abroad in London, Paris, Guatemala, India, China, and Poland, to open new semesters in Australia and Brazil, I wonder how much longer "internationalization" will be a realistic strategy for any institution. I'm on this committee because I support these plans; but suddenly any scheme that involves hundreds of people a year taking long plane trips seems doomed. I hire a contractor to renovate my house, telling myself that it's a "long-term investment"; then I wonder whether anyone will want a house in a commuter-belt town in a future when gas will cost more than champagne and Toronto, the city to which people in my town commute for work and pleasure, may be unable to feed itself. Participating on a panel at a literary festival, I give my customary response to a question about why my short stories are set in many different countries: that my peripatetic life has made me feel a little bit at home in a lot of places and completely at home nowhere; that in order to unify my personality I must be perpetually in motion. As I utter this long-held article of personal faith, for the first time it sounds pretentious and irresponsible. The next spring, as I'm lamenting the curtailing of the cross-country ski season by the premature disappearance of the snow, I read that over the March break holiday week five hundred thousand people will pass through Pearson Airport in Toronto, and I can't help but see the two events as connected.

The false consciousness characteristic of the Phony War makes us grasp at straws. Against my will, I find myself taking perverse solace in the consensus that if by the year 2040, no country in the world will be exporting food, Canada will be among the few that will still be food self-sufficient. I try to overlook the apocalyptic corollary to this common prediction: that most of the western United States will turn into a burning desert. It's hard to imagine that 150 million starving, desperate, thirsty, well-armed Americans, fleeing north to where the climate remains moist enough to support agriculture, won't make an impact on Canada's food self-sufficiency. The fracturing of the control of central governments in both Canada and the United States will mean that no one will be in a position to stem this human tide. When these gun-wielding strangers show up on the doorstep, they are more likely to take, and to take over, than to wait patiently for food handouts or blend in with local society. When I published this prediction in a magazine, a good-hearted reader wrote in to suggest that we must develop cooperative local agriculture so that we will be ready to feed the Americans when they arrive. As they will outnumber us by at least five to one, this charitable strategy is unlikely to succeed.

In a Phony War, you can't voice your deepest preoccupations because they sound like hysteria. We all live with the (mostly unspoken) knowledge of the inevitability of our death as individuals. To live with the unspoken knowledge of the inevitable death of our civilization, perhaps within three decades, is far more paralysing. Many vital activities – renovating the house, trying to write stories that will last, raising children, saving for the future, even exercising environmental responsibility – threaten to become meaningless. I remain alert for signs that the Phony War may be ending and the real war beginning. Each time a hurricane creeps farther north than it's meant to, pelting New York and Boston with weather that used to be the prerogative of Kingston, Jamaica and Havana, Cuba, I tell myself that the real war is beginning. Recently an acquaintance mentioned that she and her husband had bought five acres of land to retire on more than four hours north of Toronto. I was surprised. Economically successful West Indian immigrants in their early fifties, this couple has always expressed a preference for parts of the country where the population is racially varied. So why choose deepest, whitest, north-central Ontario....? "My husband's read the stuff on global warming," my acquaintance replied. "We have to get away from the population centres and up to where we'll be able to grow our own food." I said nothing, astonished to find someone who was acting on the evidence that surrounds us. I suspected that, like most people, I would do nothing until it was too late.

Conformist compliance with the norms of the day remains our most seductive temptation. In fact, if you are privileged enough to belong to the middle or upper classes of Western Europe, North America, or East Asia, it seems deranged not to conform. Are you really going to cash in your investments because your knowledge of climate change tells you that the stock market is bound to collapse? Life has never been so extravagantly enjoyable. Many of us may work too hard, but the recompenses are princely: great food from all over the world every day; gymnasiums, spas, saunas, and lavish night clubs in which to disperse our tensions; the ability to arrange purchases, appointments, or financial transactions, which used to consume hours of our time, at the click of a mousepad; electronic toys galore, whether for work or play; a relaxed sexual ethos that encourages experimentation with little or no moral censure and, if one takes precautions, a low level of physical risk; miraculously advanced medical and dental technology, which shrinks to mere inconveniences conditions that would have incapacitated or killed our grandparents; a booming interculturalism that spares us the ugly exclusions of the past, and allows people from many ethnic backgrounds to live, work, love, and play together; a life expectancy that has shot up by almost twenty years since the end of the Second World War; endless credit and loyalty points to encourage us to keep consuming; and – the core privilege of globalization's winners – access to affordable airline travel to almost anywhere in the world. You'd have to be a fool to renounce this feast! Even if you did renounce it, your sacrifice would make no impact. There is no deceiving oneself that one is contributing one's grain of sand to a global struggle by riding buses to work or taking vacations at a lake an hour from home when the vast tide of humanity is rushing in the opposite direction. In the most prosperous regions of Asia, Africa, and Latin America, people scramble towards a First World lifestyle which, for the first time, appears to be within reach: scramble to ensure that Pakistanis, Peruvians, and South Africans can enlarge their carbon footprints to the size of those of Americans, Canadians, and Australians by eating a globalized menu, driving cars, and taking international flights to go on vacation.

The French refer to the thirty years after 1945 as *les trente glorieuses*. Even though France was wracked by political instability, economic uncertainty, and vicious wars of decolonization in North Africa and South-East Asia, today's commentators look back on these thirty years of a progressive, cohesive, welfare state as a paradisal past to which their nation will never return. In a similar vein, prophets of the downfall of the Empire of the United States trace the beginning of the end to the 1973 oil shock and the retreat of the defeated US forces from Vietnam in 1975. Yet, if one looks back at the much-idealized stability of 1950s and the first half of the 1960s, it was a time of disciplined repression: the white picket fence, the white man in the grey flannel suit, the woman in the kitchen; the French family in its cramped apartment obedient to Papa and General de Gaulle; the Germans, models of humourless hard work as they conjured their *Wirtschaftswunder* (economic miracle) from the rubble of the Second World War; people of colour out of sight on the wrong side of the tracks; and sexual desires of all but the most socially sanctioned, married variety firmly in the closet. (We should not forget that the catalyst of the May 1968 Revolution in Paris was not imperialism overseas or the oppression of the workers at home, but a regulation that banned male university students from visiting female students in their dormitory rooms.)

Even upper-middle-class suburban houses of the 1960s in spacious, prosperous countries such as Canada, the United States, or Australia look austere by today's standards. Normally there was only one car, one television, one bathroom, tiny bedrooms, a modest array of toys for the children and four or five changes of clothing for the adults. There was no air conditioning. Today's homes, crammed with multiple televisions, computers, cellphones, microwave ovens, food processors, electronic security systems, dryers as well as the washing machines of the past, fitted with walk-in closets to store lavish personal wardrobes, equipped with an en suite bathroom in the master bedroom and at least two other bathrooms elsewhere in the house, with two or three large cars in the driveway and a garage full of discarded sports equipment, where the air conditioning hums all summer and the heating is jacked up at the first hint of an autumn cold snap, could only have belonged to millionaires thirty years ago. Our awareness of the growing gap between rich and poor in our own streets, of problems such as homelessness and our failure to integrate certain ethnic communities, of the faltering international clout of the fading former empires of Great Britain, France, and the United States, of the turbulence of the stock market, the narrow, uninspiring nature of contemporary electoral politics and the lack of interest in the democratic process shown by young people – of the precarious state, in short, of most Western societies – paradoxically obscures from view the fact that we are consuming goods, services and resources more voraciously than ever before. That, in fact, we define ourselves by consumption.

In Western societies, even the poor and marginalized must consume. Their bodies may be bloated by junk food, their minds addled by inane television programs, their ability to buy snared in extortionate lines of credit; but even they possess a wealth of luxurious gadgetry that would have made a hardworking middle-class male in a grey flannel suit gape over his white picket fence. The consumption mantra has permeated our beings so thoroughly that it is no longer permissible not to consume. Today the failure to own certain baubles inspires the open expressions of shocked disapproval that making love with someone of the same sex or a different race would have provoked in the early 1960s. I have rarely been treated with such overt disdain as when I have confessed to bright young yuppies that I do not own a car or, even more scandalously, a cellphone. Consumerist conformism is the most intransigent force in our society.

A few years ago I was teaching an introductory Spanish language class. Most of the students were well-groomed young women from upper-middle-class Toronto suburbs. Only a girl who sat alone at the back of the class had a noticeably more untamed hairdo. When she asked a question, the other young people snickered at her rural

24

accent. I noticed that she sometimes nodded off in class. Finally she came to explain to me that she was from a poor background in a hardscrabble northern Ontario town. Having come south, she and her boyfriend were putting themselves through university by working nights as cleaners in office buildings and schools. Mopping and scrubbing all night, she sometimes fell asleep in classes the next day. I gave her a reassuring, probably rather patronizing, lecture that she was doing the right thing and should persist even if it was difficult. The young woman passed the course. At the end of the semester, she said she wanted to ask me something. Her question was: "Could you recommend a good Spanish-speaking vacation spot in the Caribbean? My boyfriend and I go south for the sun twice a year and this year we'd like to go someplace where they speak Spanish."

This incident exposed many tangled assumptions about social class, particularly my assumptions and those of the other students in the course. What stood out for me was that not even someone who was struggling to get by and came from a background of relative economic deprivation, could consider foregoing the twice-annual Caribbean vacation that has become one of the defining self-validations of North American humanity. It is salutary that class barriers have crumbled to the point where this young woman could not see any reason why she should not enjoy the same vacation as her classmates, whose parents were engineers, accountants, lawyers or teachers; yet it is appalling that we do not feel human unless we consume. The consumption is paid out both in travel and in objects. I saw this again when a friend told me about his misgivings concerning the young man that his elegant, ambitious stepdaughter had chosen as her first serious boyfriend. The lad was from a troubled lower-class family. His parents had separated and lived in different provinces. One parent was on welfare; the other was marginally employed. While the girl was near to the top of her class, her boyfriend was at the bottom. Like his parents before him, he seemed unlikely to complete second-

25

ary school. One night the boy was attacked and robbed. My friend was astonished to learn that he had lost more than two thousand dollars in belongings. This young man, whose parents could give him nothing, did not leave home without carrying an iPod, a Blackberry, deluxe earphones, a case of DVDs and other toys.

Being poor in contemporary society may be to live in substandard housing, but the core of deprivation is a poverty of awareness, education, culture, sensibility, spirituality, self-confidence, and often good nutrition; it can never be, except in the case of those who are destitute and homeless, a failure to gain access to the glowing totems by which we measure our humanity. This is the one deprivation that we do not tolerate. This became clear in London in 2011, when young people who could not afford these items broke into shops by the thousands to help themselves, as if it were their right, to name-brand running shoes, televisions and sports equipment. The British middle class and their voices in the media reacted with fury. The Prime Minister commanded judges to give the looters extra-long sentences, and the judges, disdaining their responsibility to implement the letter of the law, complied. This hysterical reaction was saturated in hypocrisy. The Prime Minister, and the middle-class voters who had elected him, were the beneficiaries of an economic system that propagandized young people from birth with the credo that they would be fully human only if they possessed the items displayed in the windows of High Street shops. Consumption was essential; consumption was being and essence. If you could not enter a shop and take home one of the cherished trinkets of the day, you were not a person. In a commercialized distortion of human rights rhetoric — perhaps *the* human rights rhetoric of the twenty-first century — every screen and billboard shrieked that all individuals had the right to self-realization, instant self-gratification; that this was the paradise in which every human's acquisitive soul must find its self-expression. Once economic decline and cuts to social programs had put out of reach the constituent parts of a human soul, it was inevitable that those who felt themselves to be less than human would reclaim their humanity.

This is one of the central reasons that our responses to climate change are inadequate. It is true that in an environment where we are all deluged with information, alarmist news loses its ability to startle us into action. Even though nearly all credible scientists agree on climate change, this fact lacks the authority to induce us to move beyond our Phony War and make radical alterations to the way we live. The advance of global warming becomes one more titter of sound in the whispering backdrop of interwoven news, stock market reports, and celebrity gossip with which we are assailed. But the crux of our predicament is that cutting back consumption represents not merely a practical problem, but an existential crisis. *Emo ergo sum*: where Renaissance humans, following Descartes, defined their humanity by thinking, we define ours with the credo *I buy therefore I am*. To ask us to reduce consumption is tantamount to telling us that we must relinquish our humanity. In fact, in terms of the hyper-commercial civilization that has spread around the world since the acceleration of globalization after the end of the Cold War and the digitalization of markets in the early 1990s, this is precisely what we are being asked to do: to give up who we are and toss away that which confirms our integration and our normality; to accept that, in the terms we recognize today, we cannot be human. In order to respond effectively to climate change, we must wean ourselves of our addiction to consumption and find an alternate scale of meaning. Nobody should imagine that this will be easy. The threat of global warming confronts us with innumerable challenges: the need to cooperate, to resolve intractable disputes in democratic ways, to develop a meticulous, regulated, just allocation of shrinking resources, perhaps to legislate and enforce draconian prohibitions on driving private cars or squandering energy on air conditioning, to bring the reckless power of profit-seeking corporations under the command of elected legislatures. But the most bracing challenge is that of reimagining ourselves, our heaven, our source of meaning. Unless we can find a way not to define ourselves according to our consumption, our chances of salvaging even shards of our civilization are minimal.

CHAPTER FIVE
WHAT SURVIVES?

We are all the descendants of survivors of collapsed civilizations. Most of today's ethnic Europeans trace their ancestry to the hardy core that survived the Black Death in the fourteenth century. Indigenous Americans, from Punta Arenas, Chile to Alert in Nunavut, belong to the families of the sliver of aboriginal peoples who survived the ninety per cent decline in their numbers that occurred during the first century after the conquest of our hemisphere by Europeans. Each time that disaster has struck, a small group of people has wriggled to safety and started again.

My friend who marvelled at his stepdaughter's boyfriend's gadgetry resists my gloomy prognoses about our future. "Predictions about the future have always been wrong," he says. "Whatever people think is going to happen, you can be sure that's the one thing that won't happen! This will be wrong, too. Look around you. More people are better off than ever before. Things have never been better!"

I respond that global warming is not a prediction of a social trend or a possible future technological breakthrough; it is a precise calibration of the inevitable rise of global temperatures. In this it differs from 1960s predictions that we would soon be taking holidays on Mars or working three days a week while robots dealt with our drudge work. Most advanced civilizations have unravelled when they were on the cusp between greatness and decadence. The Classical Mayan city-states that ruled Mesoamerica from 300 AD until their collapse in about 800 AD provide clear evidence of this. The ruined city of Copán, in northern Honduras, features the longest inscribed staircase in the Mayan world. The staircase is the culminating achievement of a dynasty whose greatest king was known as Eighteen Rabbit (his Mayan name was Uaxaclajuun Ub'aah K'awiil). At the time the staircase was completed, Eighteen Rabbit had been captured and beheaded by a rival city-state; the staircase combines grandeur with decadence. The magnificent carving of the stone conceals shoddy workmanship: the fill beneath the staircase is of poorer quality than that used elsewhere in Copán. The foremen who oversaw this project were experiencing their own Phony War. Inhabitants of a powerful state, they carried on with its great plans in the face of a growing crisis. But they did their work in a perfunctory, half-hearted way. They did not build for eternity, though habit, convention and the persistence of the basic contours of social organization in the face of what archeologists have deduced to be the early stages of the famine that would end their civilization, enjoined them to continue to go about their duties.

How long will it be before our designers and builders find themselves in the same quandary as Copán foremen? And what will the future – if there is a future – think of the staircases that we leave behind? The question of what survives of a culture, and for how long, presses upon us. Led by Bishop Diego de Landa, the Spanish colonizers of the New World burned eight centuries of written Mayan literature as the work of the devil; Julius Cesar burned the library at Alexandria, possibly by mistake, an act that destroyed the definitive collection of Greek and Roman literature collected over three centuries. These two book-burners have robbed us of much of the literary culture of our American and European ancestors respectively. Yet even when cultures are not subjected to such aggressive

30

suppression, they wither over time. The buildings crumble, the paintings and murals flake, the music falls silent and is forgotten. Of the arts, literature is the most ductile: it may be transmitted by different delivery methods and even in different languages. Over the last century music has also become transmissible in a variety of forms; yet literature remains the art form that has the greatest potential to outlast the culture that produced it. We can read *The Epic of Gilgamesh* in English, French or German — languages that did not exist five thousand years ago — but how much else remains of ancient Babylon? Even the survival of a culture's literature is a hit-and-miss proposition. The pyramids suggest the greatness of ancient Egypt, the desiccated, boldly inscribed papyruses in the Egyptian Museum in Cairo provide further hints, yet there is little doubt that what has survived is an arbitrary sample that probably does not include the best of what that civilization produced.

All civilizations become ancient eventually, but the crisis caused by global warming introduces a bracing new existential anxiety to the production of art, and hence to the question of the meaning of human civilizations. Part of the appeal of art was meant to be that it was "eternal"; the values it promoted were human and "unchanging" (even though, in fact, they varied according to decade, continent, and culture) and that great art would last "forever." Under the influence of the seventeenth-century philosopher Baruch Spinoza, extreme idealists such as the Argentinian writer Jorge Luis Borges claimed that, if one immersed oneself in it, art could supplant life itself, exempting the artist from the natural cycle that included death. Yet artists, like other proponents of the humanist tradition, have been plagued by the meaning of "eternal." In a moment of depression, Virginia Woolf made the glum observation that the merest pebble on the beach would outlast Shakespeare. Even if one imagined human civilization enduring for millions of years, until the sun turned supernova, the question of art's eternity, and what this claim means, has been a source of incessant anxiety. Such visions invariably culminated in questions such as: will we preserve Homer and Shakespeare by taking them to other planets? This dilemma is older than we might suspect; generations prior to ours wrestled with the question of whether the population necessary to remember art will last long enough to make great art "eternal." Marcel Proust equated the character of the dying writer Bergotte, one of the figures who incarnates artistic creation in his seven-volume novel *In Search of Lost Time*, with a planet from which the sun's heat is retreating. In the fifth volume of his *roman-fleuve*, Proust writes: "In this way, [Bergotte] grew progressively colder, a small planet that foreshadowed the future of the big one, when, little by little, heat will withdraw from the Earth, then from life. And so the resurrection will be snuffed out, since before the artistic works of men may dazzle future generations, people must still exist. If certain animal species survive the invading cold longer than others, once there are no more people, and supposing that Bergotte's glory will have lasted until then, it will abruptly be extinguished forever. The last surviving animals will not read him...."

Today's corollary to this concern might be that when global warming floods our coasts, turns the interiors of our continents into deserts and sends the desperate billions stampeding towards milder climates in search of water and food, the most tragic loss will be that of the artifacts of Babylon and ancient Egypt (we received a foretaste of this with the sacking, in 2003 and 2011 respectively, of the museums in Bagdad and Cairo that housed many of these artifacts); the canals of Venice, the oral traditions of drowned islands in Polynesia, thousands of pagodas, temples, cathedrals, and more art galleries, concert halls, museums, libraries, and theatres than can be counted, will be wrecked by murky, briny water.

It is possible that climate change survivors will preserve literature and music – or, at least, samples of certain literary and musical traditions. Art is humanity in its concentrated essence. However horrible the billions of deaths that await us may be, humanity's most enduring loss in the long term may be that of its artistic heritage. The traditions that persist will depend on the severity of climate change. In the event of a cataclysmic rise of eight or nine degrees Celsius over 1990 levels, it is unlikely that any languages other than English and Russian will retain significant numbers of speakers, with Spanish, Swedish, Norwegian, and Finnish, and maybe Maori and Haida, existing as minority languages in the far-flung outposts of Patagonia, Scandinavia, New Zealand, and British Columbia. If the artistic heritages of Europe, Asia and Africa are lost, one could imagine the Maori or Haida carving traditions becoming the founding models for whatever future developments are feasible in the visual arts.

Even a less dramatic rise in temperature will radically rearrange the hierarchy of the world's languages. Spanish is the third most widely spoken language on the planet. Yet half of the language's speakers live in Mexico, Central America, Spain, or the southwestern United States: areas where the population could be wiped out by a global rise of even two or three degrees Celsius. The promising prospects for Patagonia mean that the Spanish of the future may be spoken with the distinctive drawl – considered eccentric by contemporary speakers – of Argentina. A similar rise in temperature could reduce to irrelevance the subtle, intricate Arabic language, currently the world's fifth most widely spoken, but many of whose speakers live in areas that are already semi-desert, or, as in the case of the extremely crowded Nile Delta, are highly vulnerable to flooding. By contrast, a moderate rise in temperature could make central Quebec into North America's breadbasket, an event that could enhance the importance of Québécois French. Currently spoken as a first language by only seven million people, its influence could grow under global warming – if the Québécois are not overwhelmed by tens of millions of marauding Americans, or subjected to the control of the Chinese, who are buying up their arable land. China is well prepared for a two- or three-degree rise in temperature; though the Chinese would lose tens of millions of people, by the late twenty-first century their position in the world, and those of their language and culture, would be unassailable.

CHAPTER SIX
THE LOSS OF THE NORTH

A few years after my trip to Alaska, I accompanied my parents to a cottage they had bought on the Ottawa River. An A-frame built on a spar of flat riverbank on the edge of a landscape dominated by the Canadian Shield, with its striated rock faces colonized by dense groves of maples and pines, interspersed with the white trunks of silver birch trees, the cottage lay on the southern fringes of northern Ontario. In this intermediate zone, where civilization began to yield to bush, I remembered the tangled abundance of the British Columbian coast and the indomitable rawness of the Alaskan wilderness. I spent the weekend staring out at the unbroken ranks of close-set trees across the Ottawa River in Quebec, my awareness of the great bulk of the sparsely populated north that bore down on the inhabited strip of southern Canada from the infinite, wild spaces at the top of the map awakened for the first time in years. Slipping into a dreamy, fatalistic mood, I spent part of the weekend writing a letter to my biologist friend. In the letter I tried to convey the sensation of feeling nearly paralysed by the human incapacity to visualize the dimensions of the Canadian north. I had not seen my friend since leaving Alaska. He spent the academic year on the US West Coast; in the summers, when I occasionally went out west, he retreated to Alaska to do his research. Our itineraries no longer coincided, but we continued to write letters. Didn't he feel, I asked, that the bush was what would finish us in the end, that the sheer uncaring savagery of the continent we presumed to inhabit would bring us down? His reply came back with a nearly audible chuckle: no, he didn't think any such thing. In fact, it was we who would finish off nature.

We did not expect this. For Canadians, at least, part of the difficulty in accepting the reality of global warming is that we always imagined we would die of cold. Winter was our monster, the equivalent in our mythology to the dragons of the ancient British Isles or the minotaurs of classical Greece. In school we read Sinclair Ross's short story "The Lamp at Noon," and other accounts of the lethal fates that awaited those who got caught outside their cabin in a blizzard. Snow and wind, as in Anne Hébert's *Kamouraska*, were the malevolent forces that kept lovers apart. We listened to radio recitations of Robert Service poems about Dan McGrew deciding to button up his shirt at sixty degrees below zero. We anthropomorphized winter until it became almost visible, nearly possessed of facial features. In her analysis of E. J. Pratt's poem *The Titanic*, Margaret Atwood summarizes Pratt's description of the iceberg that sinks the ocean liner by concluding that, "Pratt renders it as semi-alive, a sort of *Night of the Living Dead* zombie, complete with a face, a claw, a lair, and an 'impulse.'" My biologist friend did not see the ice as alive, except perhaps in terms of the microorganisms that might reside within it. In this, perhaps even more than in his scientific training, he had an advantage in grasping before I could that humans, not winter or nature in general, were the biggest danger to our planet. He did not see nature as a dragon or a zombie, but as a tapestry of interdependent life cycles that were being distorted by human intervention.

Winter was our monster, but the north, where winter went on even longer than it did in New Brunswick or Ontario or Saskatchewan, has always felt remote to Canadians who lived in the ten provinces rather than the three territories. Relatively few of us have visited Whitehorse or Yellowknife, much less Iqaluit, Resolute Bay, or Ellesmere Island. In 1967, when the pianist Glenn Gould prepared a CBC radio documentary entitled, "The Idea of North," he broadcast all of the voices against a backdrop soundtrack of the so-called Muskeg Express making its two-day journey over the rails from Winnipeg to Churchill, Manitoba. The relentless chugging of the train along the track, accompanying each interview, conveyed a sense of infinite, almost unimaginable distance. As Gould astutely observed, the north belonged to the imagination of Canadians, not, in most cases, to their lived experience. As we watch skinny polar bears churning for a foothold on dissolving ice floes on television, the spectacle feels impossibly remote; yet it does strike a chord within us. The polar bear is far away *and* he is ours. He is almost one of us, yet not quite sufficiently one of us for us to be willing to reduce our consumption in the hope that the bear will be able to tread on more solid ice floes in the future. Even if the polar bear becomes extinct, at some level he will still be there in our minds, as an emblem of Canadianness and northernness; since he was never part of our lived reality, the wound we feel from his loss will be a distant ache. The polar bear may live on in our culture as a mythological beast. Yet what will have been destroyed is the idea that Gould was able to take for granted and which he explored in his program: the belief that the north was almost infinite; that it was the place that epitomized Canadian solitude.

We now understand that, like the rest of the planet, the north has finite resources that are fast being depleted; that if one tramples lichen, it may take five hundred years to grow back. The fact of being bordered by the north has cushioned Canadians' perceptions of the world; where other countries had difficult neighbours, we had nothingness stretching to infinity. Sheer space insulated us, making us feel exempt from the crafty diplomatic manipulations enforced on peoples in more cramped geographic locations. Global warming has already evaporated this Canadian presumption of morally superior separation. Since public figures, foregoing the traditional Canadian *A mari usque ad mare* that adorns our coat of arms, began to speak of a Canada "from sea to sea to sea" – a phrase popularized in the early 1990s by the Yukon politician Audrey McLaughlin – the north has become merely one more region of Canada. The melting of the polar ice has incorporated the Arctic Ocean into our daily travails. As politicians warn of the incursions of Spanish or Japanese factory fishing ships on our Atlantic coast and the unannounced arrival of shiploads of undocumented immigrants from China or Sri Lanka on our Pacific coast, so our north is now a frontier with the world, where the opening of the northwest passage to shipping embroils us in disputes about territorial limits and Arctic sovereignty with the Russians, the Americans, the Chinese, the South Koreans, and even the peace-loving Danes. The climate of the north is heating up in all ways imaginable. Glenn Gould concluded "The Idea of North" with the prediction that in the future the north would look like everywhere else in Canada; that it would "look like suburbia." What is certain is that climate change has already integrated the north into daily life more than we ever expected. We have been stripped of a zone of purity and mystery that was one of the keys to the Canadian self-image of living at a one-step remove from the hurly-burly of planetary life. Just as we have been deprived of the knowledge that robust polar bears prowl the firm ice of our Arctic, so we are losing the defining traits of long, hard winters, the reassuring routines of consistent bird migration patterns, and the possibility of using cross-country skis for five months of the year.

This is only the beginning. There is no public will to implement a solution, and in any event it may be too late; even the most drastic measures will produce limited results in the very long term. But even if we did have the collective will to do something to save our natural world, of which we ourselves are a part, what actions could we or should we take?

CHAPTER SEVEN
WHAT TO LOVE?

The strongest human feeling is love. For centuries love was considered a noble sentiment that separated us from the animals. Love was human and unique because it inspired people to act in ways that were altruistic: a mother sacrificed her potential in order to develop that of her child, a warrior lost his life for his country. *Dulce et decorum est pro patria mori*: it is a fine, sweet thing to die for one's country. The Latin poet Horace wrote those words to cast shame on "battle-shy youths"; warmongers down the ages repeated them. The words were inscribed on the wall of the chapel of the British military academy at Sandhurst, where they remain today. During the First World War, Wilfred Owen wrote a gut-wrenching poem about being gassed in the trenches that repeated Horace's line with embittered disenchantment. "The old lie," Owen called the injunction to die for his country. Owen was dead, killed in the war, by the time his poem was published in 1920. He would never learn that by refuting Horace he had participated in changing the definition of love that would prevail in the twentieth century. Even in martial Great Britain, his poem became part of the school curriculum. People would die for their countries again during the Second World War. After this, the lust for self-sacrifice abandoned the West, absconding to Africa, Asia, and Latin America. *Mort pour la France*: the words are inscribed on plinths all over Paris. The people who died for France did so prior to 1945;

the victory of nationalist insurgents in the war of Algerian independence in 1962 illustrated that the urge to die for one's country was now stronger in the Maghreb than it was in Europe, in the former colonies than in the metropolis. The decade that followed proved that the yearning of the Vietnamese to give life to their country was stronger than that of the Americans to die for what they had been told, but often did not believe, was the interest of theirs.

As it was displaced from the centre to the peripheries of the world, this form of love changed shape, metamorphosing into a yearning to die for an idea of country, for a country that did not yet exist or had never flourished because it had been crushed by dictatorships imposed from abroad, but which was, palpably, being born. Sacrifice ceased to be a duty to an immemorial order and became an expression of brotherhood, of the mystical union of a nation that one day would live in equality. Between the 1950s and the 1970s, people died to create independent nations not only in Algeria and Vietnam, but in Cambodia, Laos, Malaysia, Congo, Kenya, Cuba, Guinea Bissau, Mozambique, Namibia, Zimbabwe, Angola, Nicaragua, El Salvador, Guatemala; in a few long-oppressed spots, such as Eritrea and East Timor, this spirit persisted into the 1990s. In the rhetoric of the independence movements, dying for the cause became a manifestation of transcendent love. Commemorating the tomb of a fallen guerrilla, the Nicaraguan poet and priest Ernesto Cardenal wrote: "Your love had a beginning but it has no end / Your atoms that were in the Nicaraguan soil / Your loving atoms that exchanged life for love / You'll see, they will become light."

With the end of the era of wars of national liberation, the urge for selfless love faltered. Owen's love, the love that spurned the notion that dying for one's country was an ennobling act, fuses altruism with egotism, combining a deep-seated humanity that decries the immorality of sending young people to die in war with the basic, selfish desire to save one's own skin. These are defining traits of late twentieth- and early twenty-first-century life. As the century progressed, the quotient of egotism surpassed that of altruism. With the onset of globalization, this self-preserving, self-exalting definition of love spread from the decadent metropolis to the newly resurgent periphery: it is difficult to imagine twenty-first-century Vietnamese or Nicaraguans choosing to die in order that their nations might live. In 2006, I visited the Museum of the Revolution in Maputo, Mozambique. Through maps, black-and-white photographs of guerrillas crouched in jungle encampments and displays of archaic-looking uniforms and rifles, this museum charts the country's armed struggle against Portuguese colonialism from the early 1960s until independence in 1975. On a cool, clear southern African winter afternoon, I coincided in the largely deserted building with a young Mozambican soldier in uniform. After we had wandered through several rooms of artifacts, the young man, who must have been born in the mid-1980s, shook his head. "It's hard to believe people actually did these things. It's hard to believe this ever happened."

Even maternal love altered its contours. Rather than devoting most of their adult lives to bringing up child after child, women sacrificed for five or six years to raise one or two children to school-age and then chafed to return to their careers. This love, which has been accused of renouncing the common good, turns out, in fact, to be working in the interests of us all. As we face a crisis of overpopulation, female impatience with endless childrearing is one of the forces that improves the odds that our species may survive. The worldly ambitions of educated women are the most effective antidote to "too many goddamn people." Consecrated in the idealistic, self-indulgent 1960s of the Western world, the idea of love yoked to an illusion of permanence began to spread. Now, rather than fuelling wars, love could stop them. Love was the alternative to carpet-bombing Cambodia; love could solve any problem, whether domestic or diplomatic. "All you need is love," crooned the Beatles. In subsequent years the global span exhibited by 1960s love contracted into a quest for exultation and personal fulfillment. Love of another became inseparable from love of self as people sought relationships that would be fulfilling, lasting; that would shield them from confronting the reality of death. Like psychedelic drugs or Asian religions, love was a far-out experience, a thrill, or, as the decadent 1970s and the self-indulgent 1980s set in, simply a Saturday-night distraction. "Love is a drug and I need to score," sang Bryan Ferry of Roxy Music.

A many-splendoured thing, love, for all its diversity, was no longer a force driving us to plough our own lives into the soil of an enriched future for our children or our country. People held off the inevitable dispersal of the love of the couple that came from having children, preserving for as long as possible their status as the unique object of another individual's affections. When the first, and often only, child came, he or she was trumpeted as evidence that the parent had transcended into a realm of privileged experience where emotions previously out of reach became accessible. Secular love was obliged to fill the abyss that was left in life with the withering of unquestioning religious belief. In a material world, the new child was an acquisition like no other; one that conferred an almost tribal respect on the parent. Often the child was conceived with the half-conscious goal of adding a spiritual dimension, a glimpse of eternity and continuity, to the lives of its mother and father. Having sought transcendence, the new parents then complained about the mundane inconveniences inherent to parenthood, the drab ordinariness that this form of sacrifice lent to daily life.

It is this vision of love, and not those that preceded it, that separates us from the animals.

The female cat or rabbit or bear produces offspring at every biological opportunity and defends them selflessly. Hikers in remote areas know that black bears ramble off into the undergrowth to avoid people, unless their young are present; then, and usually only then, they become dangerous. Self-abnegation is not confined to the maternal instincts of the female. When confronted by an external threat, the wolf pack's dominant male dies without thinking, because his instinct is to fight for the survival and continuation of his followers and his kin. The wolf resembles Horace far more closely than he resembles Wilfrid Owen. The wolf would understand Horace; he would not understand the Beatles, or the busy, acquisitive yuppie who seeks self-realization through cautious late parenthood. Prior to the twentieth century, when love – a word that often mingled sacred and profane devotions – was asserted to be that which separated us from the animals, our conception of love was in fact more instinctual and natural – more like that of the animals – than it is today. In the late nineteenth century, as technological society spread, extending the networks of factories and railways on which advanced economies depended, developing the mass-produced automobile, the ocean liner and the first airplanes, and culminating in the Anglo-German naval race that set the scene for the First World War, writers such as D. H. Lawrence dissented by exalting the raw power of the instinctual over the civilized, of nature over technology, of sex over social convention. Yet, by this time, the call for a return to instinct had become retrograde. If Lawrence, who died at the age of forty-four in 1930, had lived into the next decade, there is little doubt that he would have been tempted by the appeal of fascism. In a similar vein, Albert Camus, who praised the natural world of the North African coast, its sun and sand and spray, both in his fiction and in lyrical essay collections such as *Noces*, was on course, at the end of his curtailed life – he died in a car accident at forty-six in 1960 – for a reconciliation with the Catholicism he had put behind him during the Second World War, when he became one of the originators of French Existentialism. The independence of his native Algeria, which was formalized two years after his death, would have pressured him to adopt a conception of love hostile to that of the revolutionaries. If he had lived to see the May 1968 Paris student revolts, Camus, like André Malraux – another 1930s and 1940s leftist – almost certainly would have marched in support of General de Gaulle.

To save ourselves from the damage we have done to our habitat, we must love our species. We must love it as if the whole species belonged to our pack, with an instinctive, wolf-like atavism; yet returns to instinct are always retrograde. In the post-1968 period it has become obvious that attempts to reinstate "traditional values" – particularly in the name of unbending interpretations of religious faiths – always oppress the female. Without educated, active women who have an interest in controlling their fertility in order to experience motherhood as an element of a diverse life, rather than as the reality that annihilates all other experiences of life, human beings will not survive. The love of nature and instinct as a refuge from alienating technology, praised, in different ways, by Lawrence or Camus, no longer makes sense. Today, without technological advances, we will not spare enough of nature to thrive as the mammals that we are. Technology – cars and planes and oil extraction and overgrown, unmanageable cities, medical advances that lower infant mortality and prolong life, bio-foods that both enhance production and eradicate biodiversity – is the origin of our dilemma and the only possible solution to it. Love, which was once conceived as a perception of our fellow humans through the veil of spiritual definitions of worth, cannot persist as egotistical self-realization. Nor can it flourish as a return – another retrograde quest to resurrect the past! – to the idealistic yet ineffectual search for global love that inspired the Western anti-war movement of the 1960s and 1970s. Love cannot be stunted into survivalist self-preservation: a stockpiling of canned food and weapons to save one's family when the Hobbesian "war of all against all" sets in. The most unimaginable mental evolution that we will have to make to ensure that the hundred thousand years of human civilization of which we are the inheritors is not lost, is the transformation of love. This will be much harder than learning to grow vegetables or to live in one place, deprived of business trips to Tokyo or Singapore, or vacations in Paris or the Seychelles.

Love of one's children or grandchildren has been evoked by innumerable tyrants to induce young men to go to war for the cause of the nation or the regime. "We must sacrifice for future generations," Stalinist propaganda proclaimed. The individual must die so that his children can live in the thousand-year Reich, in a Europe free of Nazism, in the perfect Communist society, in a world free of Communism, in a secure Jewish state, in a land free of imperialism, free of terrorism, or free of infidels. The twentieth century exhausted the credibility of such martial claims to love. Both in East and West, the rot had set in by the 1960s as hippies and refuseniks exercised the force of their pacificism; in the quarter-century since 1989, the rejection of self-sacrificing love has spread to countries that were forged in revolutionary struggle, and to all corners of the Middle East. No one believes any more in reducing the quality of our life so that our children and grandchildren can live in a better world. In fact, we insist on the opposite: higher living standards so that the new middle classes of India, China, Brazil, South Africa, and twenty other countries can live like the middle classes of Dallas, Toronto, or Frankfurt.

Yet, as never before, this is what we must sacrifice. We must curtail consumption by reducing our pleasures. The call to sacrifice for a better human future has never been so authentic or urgent, and it has never had less persuasive power. Canadian Prime Minister Stephen Harper, a snarling opponent of policies to address climate change, scoffs when it is suggested that he would implement such measures if he loved his children. Yet, if we cannot think of our immediate families, we must, as corny as it sounds to cynical, know-it-all twenty-first-century ears, think of the human family. We do not need to save the planet – the planet is safe for millions of years to come – but we do need to, and may just be able to, save human life and preserve some of its extraordinary legacy. We need that trickle of survivors that enables humanity to start again. We cannot know what form this starting again will take because this will depend on the nature of the catastrophe that awaits us in 2030 or 2040. We cannot predict the contours of this new world, its political arrangements, what sort of technology it will be able to employ. James Martin, like the early twentieth-century science fiction writer Olaf Stapleton, predicts the rise of Patagonia, the only stretch of land in the southern hemisphere that will retain a climate capable of supporting agriculture, as the dominant base of urban populations in the mid- to late twenty-first century. "Patagonia," he writes, "perhaps the most beautiful place on Earth, will be covered in wildflowers and have climate-change cities."

As we move to the far north or the far south, we will need to be governed by an ideology of cooperation. The late twentieth-century recasting of love as self-fulfillment and egotism evolved into one of the necessary conditions for founding the business-driven free-market societies of the early twenty-first century. These societies eroded the egalitarian welfare states that had been created in the northern hemisphere since the 1940s. The exaltation of self morphed into the inalienable right to engage in any business practice that one pleased. The traditional democratic pillars of freedom of speech, freedom of thought, and freedom of assembly suddenly became mere adjuncts to the all-powerful doctrine of "free enterprise." When, at a public meeting, my father spoke against the construction of an enormous shopping centre in the middle of his residential community, an acquaintance in the audience shouted in horror: "But that's against free enterprise!" The restructuring of Western societies since the 1970s, in ways that have exempted large corporations from taxes, replaced the political stories that dominated serious newspapers as recently as the 1980s with "business news" about the latest lurch in the stock market (or "lifestyle" stories about the rich or famous). These trends skewed the distribution of wealth by eliminating the social programs that enabled the middle class to flourish (by 2009, 3.8 per cent of Canadian families controlled 67 per cent of the country's financial wealth), encouraged companies to merge with and take over others until the tyrant's cry of necessity, hurled down on the masses from the CEO's corner office, became the central force that influenced government policy.

These changes rendered any plea for egalitarianism or cooperation suspect, a sign of weakness rather than strength. Intelligence, realism, even personal worth within society, have been signalled by bowing one's head before the dominance of corporate logic. Just as the call for self-sacrifice has never had so little persuasive power, never have ideologies of social cooperation, no matter how mild, been so marginalized. The culture of Facebook "personal settings," of social networks consisting of "virtual friends" who do not forge deep or lasting bonds, dilutes community. These technologies may be useful for spreading news and coordinating political action among co-religionists, yet they mitigate against the kind of lived community that is essential to cooperation in times of dire crisis. Our personal communications technology has heightened social fragmentation in a way that is more about isolation and self-absorption than it is about an individualism that might promote self-sufficiency. As tools for democratic debate – for reaching consensus about what to do in a crisis – on-line social networks are ineffectual. Extensive research has shown that such networks may be used to coordinate activity among like-minded people – to assemble young Egyptians in Tahrir Square, to permit looters to converge on an undefended London shopping district while the police are occupied elsewhere – but that no one ever changes his or her mind as the result of an on-line conversation. Most people visit only the sites of those with whom they are in agreement; when they stumble across a view that runs counter to their beliefs, they click away from it with a mental shake of their head. In other words, there is no such thing as on-line democracy, effective on-line debate, on-line social problem solving. Feelings of common purpose may develop between on-line "friends," but they do not summon up the sense of commitment that one feels to people with whom one has shared lived, rather than virtual, experiences. As more of our professional and social activity migrates into the virtual sphere, the possibilities for coordinated activity dwindle. The bonds of community exert only feeble claims on most of us. Jean-Jacques Rousseau's vision of the sexual energy of the couple radiating out to bathe the wider community in an essence of caring has been negated by the recasting of love as self-fulfillment. In the absence of ways in which to work together, we return to the Hobbesian world where we fight in the streets over the last stockpile of canned tuna. It is a good thing for humanity that the love praised by Horace – the love of automatons eager to march off to war – has nearly vanished from the modern world. But in the absence of an alternate version of transcendence, of a love that gives us access to a vital, natural connection with the rest of humanity and with our vanishing natural world, it is unlikely that we will scale down our consumption to meet the available resources by any means other than killing each other.

ENVOI

On a long summer evening, as I ride a train across northern Germany from Berlin to Hamburg, I think about the tall trees of the British Columbian coast and the future that threatens them. In a model of environmental responsibility, the train has filled with well-dressed men and women who leave their Mercedes and BMWs at home to make their business trips by rail. Outside the window, every field is cultivated. Ranks of solar panels and wind turbines, interspersed among the ripe hay and ruminating dairy cattle, stretch to the horizon. The train slices swiftly through the countryside and enters the city: Hamburg, a major industrial centre, has little suburban sprawl. I remember equivalent scenes close to my home in southern Ontario: one person per vehicle; the SUVs and huge sedans boxed in one behind the next for tens of kilometres along Highway 401, stopping and starting as they nudge through the clogged suburbia that has engulfed Toronto; the remaining fields scarred with senseless outbreaks of new housing that gobble up the best arable land in Canada, neighbourhoods whose residents have no choice but to drive cars; no evidence anywhere of an attempt to rationalize transport or generate alternate forms of energy. The difference between those areas of the world that have begun to respond to the demands of climate change and those that resist responding is obvious and humbling. The tragedy of our time is that, in the long run, the difference may not matter.

ACKNOWLEDGEMENTS

Above all, I thank Ana Lorena Leija, whose research into climate change obliged me to think about the subject in detail. I'm grateful to Steve Osborne and Mary Schendlinger at *Geist* for support over the years, and for publishing the column that was the germ of this essay, to Juergen Boden and Sander Jain for persuading me to write this piece, and to Daniel F. Doak, who experienced some of these events long ago and returned three decades later to offer valuable comments on the manuscript.

.